BIM-KIT

THE PRACTICAL GUIDE TO BIM CONSTRUCTION ADOPTION

AUTHOR: JERO JUUJÄRVI

Copyright © 2020 by Infrapioneer Group.

Learn best practices from the top influencers in BIM at:

infrapioneer.com

Why I Wrote This Book

BIM this, BIM that. BIM has become a hot topic at coffee tables. I wrote this book to guide you on the right path and to not get overwhelmed with overflowing information around BIM.

These practical methods I am describing in this book have helped me multiple times to find the accurate and most beneficial targets, scopes and plans for execution.

Even for BIM industry veterans, they have good processes and focus on certain area of BIM adoption, but when the goal is to be able to replicate the value, people get crippled by requirements and guidelines.

I have found this book's approach most convenient to maximize the benefit and minize the confusion.

Why You Should Read This Book

How do I implement BIM? How can I improve BIM adoption in my projects? How should I approach BIM to make it effective?

If you are looking for a step-by-step guide to overcome the mountain of information called BIM, this is the best option.

You will learn what we need to do to implement BIM successfully and more importantly grow and improve your effectiveness and value gained from BIM-based production.

This book will help you pull all nuances your business into one clean and adoptable practice that will help your whole organization to maximize the benefit and down-size your risks in BIM.

Table of Contents

Why I Wrote This Book

Why You Should Read This Book

Table of Contents

Chapter 1. The Fundamentals

Chapter 2. BIM Value

Chapter 3. Deploy, Simplified

Chapter 4. Afterword

About The Author

Chapter 1. The Fundamentals

Before we go into actionable practices, let's look at this book's concept from the big picture.

The goal in BIM-KIT is to build a development machine. A pro-active routine to understand underlying issues and desires, finding solutions to achieve the goals and making sure it is well deployed and supported.

To put this simple: Development Machine is "the process" of improvement and you are the facilitator of the process in the organization.

And the steps are quite simple and straightforward.

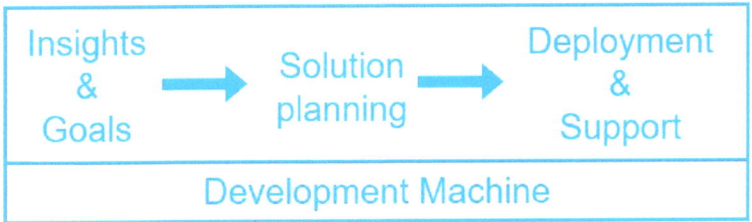

You cannot use the machine alone; it is meant to be deployed into multi-level, multi-team organizational structures.

It is also not deployable only to management level or team level. It is designed to be deployed organization-wide and thus takes enormous effort to be deployed.

This enormous effort is your first challenge, but after it is deployed, it becomes a routine check every year or a half and runs very smoothly providing valuable intel and speed up your company's adaptational capabilities.

INSIGHTS & GOALS

This part is the BIM Value chapter's recipe.

Your aim is to facilitate meetings where you gather insights on issues in projects. Issues itself do not provide many insights, so you will dig deeper together to find the root causes of the issues.

When Root causes are clarified, you will take a step back and ask "What then, if all would be perfect? How would the work look like?"

Gather up their ideals that you can convert into goals.

Now when goals and root causes have been discovered, you can start thinking possible action steps to achieve goals and solve the root causes.

This itself is very effective and useful, but we are not looking for smaller benefits, we should aim to move the needle the most with the least effort.

> Important here is to notice the flow of a meeting. You should be able to facilitate and replicate the meetings and teach others to facilitate similar meetings.

A good way to start is to facilitate these types of meetings with the heads of units and their project managers.

After these meetings are done, the heads of units should facilitate similar meetings to each team in their unit.

The flowchart would look something like this:

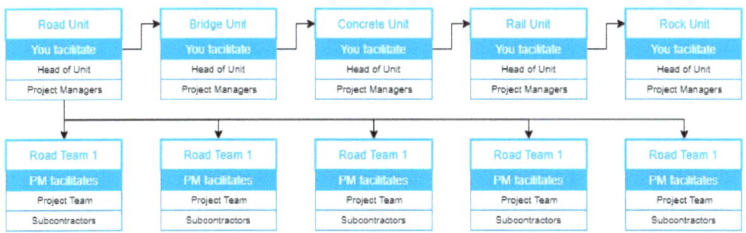

At this point, your goal is not to start planning solutions. Just gather insights and teach how to facilitate this for teams in an organization. More data, better actions taken afterwards.

SOLUTION PLANNING

So now you should have a large number of insights gathered from the previous section, it's time to start planning solutions!

But, before we do that, there is one step to do first: concatenate the insights.

Why? Because we still aim for big results. I am sure many heads of units or project managers have told you many specific issues and action steps to be made immediately.

Even though it is fun to start building up solutions such as conducting seminars, training sessions, or buying tools and services, we should do it smartly.

To do this part smartly, you should look for similarities on issues, ideals or action steps and form workgroups around them.

Let's start with the first task: Look for similarities.

Simples way to concatenate the insights is to list all issues, root causes, ideals, and action steps into an excel sheet and build visual graphs to see similarities

List all topics of insight in the vertical line and department/team units on the horizontal line who have the same pain.

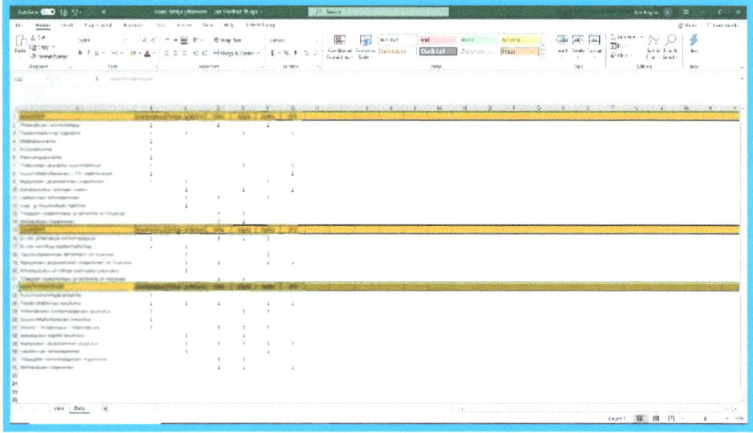

When that is done, you can create a new tab and insert graphs of stacked 100% bars in it. This way, it becomes visually clear to see if there is any synergy between different units to collaborate and plan the solution.

When you have this clear indication of possible collaborations, it is time to move to step 2.

Start facilitating workgroups between different units or teams inside one unit. The goal of these workgroups is to talk the issues/ideals out and plan action steps toward a solution that will benefit the biggest amount of people in the company.

This way, you are doing two things:

1. Building a solution that benefits most of the individuals in need of this solution.
2. Collaborating with many experienced people from different roles, allowing bigger insights to be gathered to plan a perfect solution for an issue

DEPLOYMENT & SUPPORT

Now you should have all workgroups active and you, as a facilitator, are pushing it forward to the end results.

As an end result, you should now have action steps to achieve the desired solution and now is the time to deploy it into the organization and supporting the colleagues to learn the new way.

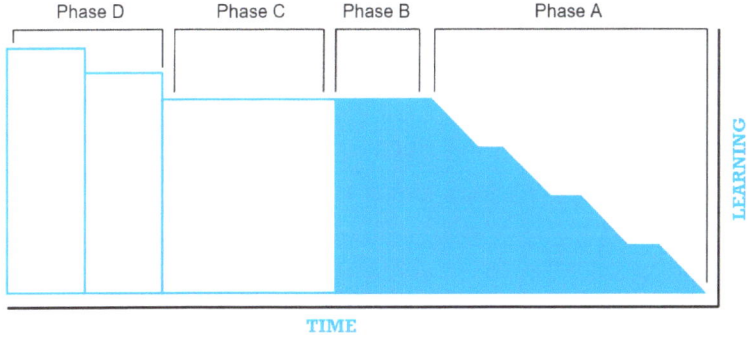

To deploy and support, I have found the following way to be most effective:

A) Teach small bite-sized pieces that your colleagues can take into immediate use and benefit from it (Instant gratification) while you are doing all the management below the surface. Your colleagues should not worry about information management, data quality, purchase and activations or anything extra. They should solely focus on those bite-sized pieces while realizing the benefits of using this new way.

B) Now you have achieved the benefits of the team utilizing the new system effectively. But before moving into the next phase, we should have a "breathing break" so that all slower learners reach the same utilization and benefits gained from the new solution.

C) Now is the nightmare time for the team: Responsibilities! At this point (usually 1-2 months after starting the learning), it is time to set clear responsibilities for each person or role. The things that you had to do in Phase A and B are now delegated to the team. It is now the team's responsibility to take good care of information management, naming, data management, and quality management etc.

As you can guess, they will fail. But failure teaches them. When you drop them first time into the water, their only choice is to learn to swim. Let them fail. After failure, they realize the importance of maintenance behind the solution and start naturally maintaining it to get the benefits they got earlier.

D) You should not move to Phase D before they can maintain the current requirements. If they can maintain the current requirements, it is time for you to start stacking up new learnings. This is slow learning, but it happens when you are not there. Basically, you visit them every 2-3 months and teach them something new to the solution. Then let them take proactive actions to take it into the field and get benefits.

Proactive action may not always occur, so you may need to take a couple of days to be with them and support them to adopt the new system, but you should not be in an active role to achieve it.

Phase D is everlasting. The Solution that you planned out and deployed gets lots of new feedback, ideas, and problems. Your task is to observe, absorb and teach it to other teams. This is good continuous teaching to improve from the existing level.

Now you have the overlay of the working process of BIM-KIT. In following chapters I take a deep dive into specifics so prepare your pen & paper!

Chapter 2. BIM Value

In this chapter is we delve into how you can start utilizing BIM and benefit from it by doing your work faster and make your business more profitable.

BIM is the new hot potato in the industry and everyone is talking about it. But before we go ahead, I want to reinforce that BIM can be very misleading as well.

BIM isn't a new format for PDFs and papers. BIM isn't something that only architects and civil engineers design and BIM coordinators use.

BIM is a tool. A mentality. An approach. To connect, to enhance, to improve.

BIM is not just a visual aid or project data, but a stairway to higher efficiency and guidance.

This first chapter will take you on a journey to explore the possibilities of utilizing BIM for improving your work and the overall success of your project.

Map Out The Ideal Project

"BIM is the Future Construction." I hear that very often these days. Those who say that speak as if BIM is a magic wand that will convert failure into success, and magically bring home the +20% savings it promises to bring, while they sit back and relax.

> But what they say is just half of the truth. This is how I see it: BIM can bring you +20% savings, only if you know how to use it. After all, BIM is just a tool. If you don't know how to use it, it will only lead to injury"

So, let's have a little brainstorming. Paint your picture so to speak. So open up your word or pen & paper and let's start.

Task 1

Let's imagine. All the people in your projects have all the required knowledge to do their part of the project quickly and with top quality. What would it look like?

> START WRITING

How would that affect the resources used?

For example: if foremen would always know what each excavator and workmen are doing and on what quality, do we need the surveyor that much?

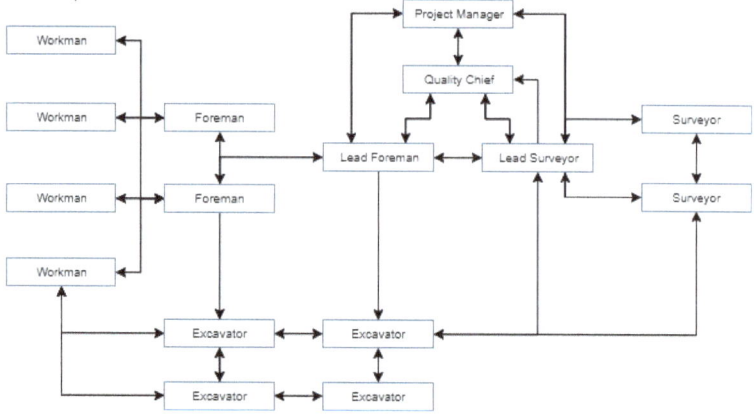

Or if the excavator drivers would know what to do, and how to review the quality of their work, do we need the foremen on-site anymore?'

What resources are we going to save if everyone on site has all the information at their disposal?

Let's look a bit further. What kind of information from the previous phase would be most beneficial for you in your work? How should it be sorted?

How about the next phase after your work? How should the delivery documents look like to get the most value out of the next phase?

> This may not be in your interest, but earlier or later they start demanding it, so why not try to look things on both ends.

In an ideal world, would the work that you described ever look like that? Or did you miss some points on how the projects would get done if everyone would have real-time data they need? What else would happen if everyone would have all the information they needed.

I don't talk about BIM models, machine guidance systems, visualization models or anything. These all are tools to achieve the ideal goal. The Ideal goal is what you have described above.

To put this simple: The goal of this part is to think about how to achieve the projects as efficiently as possible. Don't think the solutions yet, only what it would look like. We get back to this later.

Root Causes

Before we can move on, we need to discover another part: inefficiencies.

A project becomes effective when you do this:

allow the flow of valuable information to all phases of infrastructure, while at the same time stop the flow of non-valuable information.

You must know at each part of the project what information the individuals need and in what format, to make that phase as efficient as possible.

Let's move on to the second part to find current issues and the root causes that create these issues. Still, we are not looking for solutions, just clarifying the building parts for ourselves to build the solutions on.

Task 2

So let's start with the problems. List all of the problems that you and your colleagues have encountered in your projects.

These can be: team spirit, learning new things, results from surveyor taking too long or the excavators had already started to build the next layer, etc. It can be anything that slows the process, builds up the gap of information or not letting people work due to a lack of information.

START WRITING

Alright, you got a good list of problems in your projects. What now?

Well, the problem stems from somewhere. We need to find the root cause of these problems.

The easiest way is to ask why until you find the cause that creates or enables the problem in the first place.

Problem: Communicating the possible risks

#1 Why: Why your foremen are not communicating possible risks?
Answer: We didn't saw the mistake happening.

#2 Why: Why you didn't see the mistake happening?
Answer: We were not in that place during that time.

#3 Why: Why you weren't in that place
Answer: We had other work to supervise elsewhere.

Result: Not enough resources to supervise all details of the project

My example was very simple and straightforward. Don't worry about trying to solve the problems at this stage. Just try to find their root cause. Some like this one are simple, others may require more digging.

PROBLEM SOLVER

Great, now we have mapped out the problem's root causes, and the ideal work situation.

The next step is to look at both and find ways to solve problems to achieve ideal work.

Let's take the previous example:

Problem's root cause: Not enough resources to supervise all details of the project

Ideal work situation: Foremen would always have knowledge of what each workman and excavator is doing and on what quality.

TASK 3

Let's break down the Ideal situation into pieces first:

1. Automated quality assurance
2. Information about where each workman and excavator are and what they are doing

Now, we can start exploring possible solutions (tools, software, new process, new protocols) to solve the issue.

> You should not re-create the wheel from scratch if someone else has already done it. Collaborate with your unit or even at the management level to plan common practices that enable you to achieve your ideal project.

Write up all the root causes, their ideal equivalent and then a solution

Great work!

This is BIM! It's the utilization of visual and digital to serve the project most well and help each individual to succeed.

Using the same system, we should dig deeper and do benefit-specific discovery.

I highly recommend bringing your team or leaders together and brainstorming with them.

Like earlier, let's follow the same format

- Map out the ideal level of operation
- Find the deep root causes
- Think solutions

Here are few easy benefits you should dig through:

- Information Management that servers it's users
- Information that Allows Collaboration and Support
- Quality Assurance and Delivery Material ready when the task is completed
- Getting Tasks Done Faster
- Knowledge and Sharing / Training New Practices

Good job!

This method is usable at all levels. We can go through this with each project manager, but the project manager can also repeat this process with each role assigned to the project.

For example, for a construction project, I would also interview foremen, surveyors, quality chief, site engineers, workmen, excavator drivers, etc.

Basically, you will interview all the people working directly under you or the ones sub-contracted. The more information you get, the better the insights you will have at your disposal to build solutions that provide bigger results.

So do that next following the same format.

CHAPTER 2 DONE!

So what's next?

Well, if we look at your notes, we can start to see clear notes on how to support project members on issues and drive them forward into an ideal work situation.

I hope you had good sessions with your colleagues to plan out the suitable solutions to increase efficiency. With BIM or not!

Chapter 3. Deploy, Simplified

In this chapter, we will progress from insights to practices.

It will teach you to start building an action plan for the execution and deployment of BIM.

You are the change agent and you need to get the process working with BIM enabled tools, otherwise, there will be a lot of chaos and disappointment.

It is critical to plan this through because you will only get one chance to convince your project team of its effectiveness.

Enjoy!

Before You Cross This Line

Before diving into the details, let's start with the pre-requirements so you can get the most benefit out of this chapter.

Firstly, you need to have gathered insights on the company's operations. You need to understand the underlying root causes, ideals and potential solutions that will aim to support your teams to succeed better.

I highly recommend finishing your discoveries in previous chapter and do the heavy lifting of gathering insights before moving to the planning phase.

Understanding these insights will allow you to create the steps for better deployment in a much more effective and efficient manner. You will have the ability to focus on all different corners of a problem and tailoring the solutions that will work in current processes while also supporting everyone and not making it a pain for any of the team members.

Before we proceed, have these listed (these are found at previous chapter)

- All root causes + similarities between different units?
- An ideal level of operations (Where each unit is aiming to be)
- Potential solutions

> Make sure the potential solutions are approved & supported by the department's leadership.

BUILDING THE BIM PLANS

All good from the previous chapter? Excellent!

So as I spoke in the previous chapter, you need to know the root causes, ideals and solutions.

Now, it's action time.

Start writing BIM plan documents for your operations. BIM plan documents are those documents that people who are not part of the operations or well versed with BIM-based production understand.

The goal of these documents is to clarify our approach with BIM and give a manifestation for our final results. It's a neat document for the clients and subcontractors to understand how they can benefit from these practices and what the end results would look like.

The results are:

- Requirements for own work crew, so everyone is operating under the same standards. This includes file structure format, file formats, file naming format, model code format etc.
- Subcontractors know how you want them to operate
- Client's know how you want to prove quality to them

This puts you in control of your projects. If you are a subcontractor, these documents probably are already existing, but you still have time to input your own thoughts into these documents after the contract is awarded and before the commencement of the project.

The ultimate goal is that everyone understands how we do things on the project site.

I would like to break it down as follows:

MASTER PLAN

- Master BIM Plan (MBP)

FOCUSED PLAN (ATTACHMENTS OF MBP)

- BIM-based production (BBP) quality assurance plan
- BBP Information management plan
- BBP Surveying work requirements
- BBP Modeling work requirements or link to national/global standards of modeling

I know, it's brutal, but in the end, we will end up with plenty of different documents for different purposes. This will avoid the rest of the workforce to feel BIM as a monster and making BIM more accessible. So you are making everything a hundred times easier if you start creating and using these documents in your company.

TASK 1: MBP (MASTER BIM PLAN)

So the goal of this document is to touch upon general topics about BBP (BIM-based production) and link to focused documents.

Many people who are not directly part of BBP operations usually get this document and won't feel a need to read any of the focused documents, so it should be the general handbook for them on how you want things to go.

Include in your MBP at least the following things:

1. **INTRODUCTION**
    a. Goal of BBP
    b. Contact personnel of BBP Project
    c. Roles and Responsibilities (table of responsibilities and who operates, who checks and who is responsible for it all. Each responsibility can have many personnel, but preferably only one person is responsible for overseeing it all.
    d. BBP Meeting standards & communication
    e. BBP Modeling requirements (mention here national/global standards or link to your document)
    f. Project coordinates & control measurement basis
    g. BBP software used in the project
2. **INFORMATION MANAGEMENT** (link to the information management document)
    a. The softwares used & their benefits
    b. Documentation practices
    c. Folder structure + naming standards
3. **BBP QUALITY ASSURANCE**
    a. How the earthworks quality assurance happens
    b. How to rigid structures (Bridge, Foundation, etc.) quality assurance happens
    c. DWG/DXF Maps
    d. General quality assurance (Machine guidance control measurements, reporting, calibration of surveying equipment, etc.)
4. **BBP PROCESS**
    a. Induction & Work planning
    b. Machine guidance systems
    c. Drones

d. Frameworks and Staging
   e. GNSS-Receivers, tablets, and usage of BIM models & background maps
5. **BBP DIGITAL DELIVERY**
   a. Deliveries content & scope
6. Any other general aspect you want to declare here, remember what's for the document exist. Not all nuances need to be declared here.

TASK 2: BBP QA (QUALITY ASSURANCE PLAN)

In this document, you will go into specific details of the softwares, tools and work methods being used in doing the quality assurance of your project.

Basically, this document should convince your client that you are documenting the project in accordance with the client's quality standards and have control measurements & practices in place for ensuring its perfect execution.

Some of the aspects to be included in this document are as follows:

- How does machine guidance work?
- What kind of control practices do we put in place to guarantee machine guidance's quality?
- What is the accuracy tolerances for each earth layer that we are cutting/filling?
- Mechanisms for quality assurance and methods of dealing with bad quality issues.
- What softwares are being used to perform quality assurance?
- Methodologies and approaches to be used for performing quality assessment in softwares.

- Mechanism of documenting all the aforementioned information.

Apart from that, feel free to add other important information that will clarify your approach of proceeding forward with the project. The point to remember is that this document is not about telling the step-by-step approach but an overall process of doing quality assurance to the client.

TASK 3: BBP IM (INFORMATION MANAGEMENT PLAN)

In this document, your goal is to detail how you will manage the information of a project (quality, process, designs, reports and other)

At a minimum, I would include the following things in the document:

- The folder structure of a project bank
- File naming logic
- File formats
- Process of adding/revisioning documents and keeping the right people informed about updates.
- The roles and responsibilities of each individual in information management

This document is not only crucial for quality chief, surveyor or client. This is the foundation for your projects communication and collaboration. Not any less important document than any other! Do it well!

TASK 4: BBP SWR (SURVEYING WORK REQUIREMENTS)

In this document, you are telling the direct requirements that the surveyors need to know about doing the project.

Few things to be included in this document for the surveyors are as follows:

- Coordinates & elevation system.
- The requirement from the surveyors "Surveying Plan" when the project starts.
- The softwares they need to use.
- The code libraries to be used for logpoints (nomenclature).
- The type of control measurements (Machine guidance systems, Designs etc.).
- The standards for making models for machine guidance, if required.
- Any other requirements from the surveyors for supporting the rest of the project team

Surveying and Quality goes hand-in-hand in infrastructure construction and Foreman should be fully aware of their work. Industry has had a phenomenon of outsourcing surveying work, and thus this document is crucial to get your work crew moving same direction.

Task 5: BBP MG (Modeling Guidelines)

In this document, you will outline each exact requirement of what is to be included in the models. Usually, it is quite an enormous task to do so and, therefore, I highly recommend researching global or national standards for modeling for earthwork and modeling for rigid structures (IFC).

Upholding these standards will not only help you get better models but will also make them work with the softwares in a more seamless manner.

Share The Manuals

Now that we have created the guidelines and figured out the general high-level way of operations and how we want to run our processes, we can come one level down from there to the manuals.

The common problem with high-level documents is that it is TOO HIGH LEVEL. People need practical step-by-step guides & training to learn new habits. And this is why this part is very critical.

Manuals will support you to get the workforce utilizing new work methods that will benefit a wider group.

Here is a format I used and have found it effective:

1. Directory Page/Table of Contents to jump to the right page of the manual
2. Chapter titles: Task name + benefit from it
3. Chapter description: Write full benefit and how it benefits others by doing things this way
4. Lots of pictures depicting workflows (less text, more pictures).
5. Step by Step guide on using the tool/feature sets

When creating manuals, I highly recommend making one manual for each role and one manual for each software or tool.

For example, the CAD project bank software I have made includes the following manuals:

- Foreman's guide (how to view data and utilize in their GNSS-receiver)

- Surveyor's guide (how to validate data and check for bad quality)
- Information manager's guide (how to manage the information in that software)
- Excavator driver's guide (how to validate own logpoints and fix if bad quality)
- Client's guide (how to view quality documents and approve or complain about them)

Make tool based manuals, task-based manuals and, always, role-based.

Each one should contain step by step guides.

Keep it simple and easy to follow. A lot of pictures will help you greatly with the task. These manuals are very effective support tools to get your work teams to adopt new ways of doing work.

Training & Support

While manuals help a great deal to utilize new habits, we are still in a very traditional industry.

Therefore, face-to-face communication is still the most effective means of making a change. Manuals are there for guidance in the absence of an expert.

Thus, training & support is your doorway to push new things into the sites and get people to adopt them and maybe even praise and convince others of BIM's value.

So go and schedule training sessions for each new project your company wins. Arrange your calendar in such a way that you can visit the site at least once a month inquiring how everybody is doing and knowing if they had any challenges regarding BIM-based production practices.

I may not be the best teacher in the world and thus my tips for teaching are limited. However, I have found that the best way to learn something is by doing it yourself.

Also, as a hint of wisdom, I would like to quote the words of a good friend of mine,

"You don't know the tool if you can't teach them forward"

I think this summarizes it all. You need to be an expert. You need to know the tools to be able to teach them to others. So be that guy for them.

SET UP IN THE PRACTICES

To finalize this chapter, I think we need to set these things in order.

If you have produced all the documents and do the least effort possible by opening an email, adding all 20 documents as attachments and sending them to your project team, you have failed immediately.

Therefore, I want you to do the following:

Create a process that covers how are you going to approach projects that are launching soon to learn and adopt the practices detailed in your documents the most efficient way possible.

We do not want to overflood them with information. They need to be bite-sized and in situations where they can they them immediately into use and benefit from them.

The goal of building an operating plan is to make your work duplicable. So if a new BIM champion arises from your company or is hired in, he has a good start to begin pushing BIM to your projects.

It also prevents you from overloading people with irrelevant information.

The key to good information management is:

Giving the right information to the right people at the right time

So do that!

Chapter 4. Afterword

This book is meant to be the foundation for your organization's BIM journey. If you get all of these things correct, future iterations and improvements will be much less of a challenge.

I see BIM as the first step to digitalization, automation and robotics. What you are, right now, taking with you and starting to implement is the start for future development.

We are basically converting real-life effort into digital goods that softwares and robots in future can utilize with much higher efficiency.

BIM is just a start. It is important to keep building the base for your future projects so you have the infrastructure of your own to keep improving work in the future as fast as possible.

To dangle a carrot little more, there is a way to make it even more effective and faster, but that's the story of another time.

If you are interested, reach me out at jero@juujarvi.co and we can take this conversation even further.

About The Author

JERO JUUJÄRVI is vibrant life-long learner who has explored and worked in many industries and many professions. Adamant desire to understand all nuances and transform it into beneficial and repeatable practices is at the core of his workings.

Learn more about Jero at infrapioneer.com

ONE LAST THING…

If you enjoyed this book or found it useful I'd be very grateful if you'd post a short review on Amazon. Your support really does make a difference and I read all the reviews personally so I can get your feedback and make this book even better.

If you'd like to leave a review then all you need to do is click the review link on this book's page on Amazon here: http://amzn.to/yourlink (direct link to the "Create a review" page on Amazon for your book – you can't get the link until you upload your book first. Then, when your book is live, go get the link, insert it, and re-upload your book.)

Thanks again for your support!